TINY CACTUS PUBLISHING

COPYRIGHT 2017 ADULT COLORING BOOK BY TINY CACTUS PUBLISHING.

ALL RIGHT RESERVED.

NO PART OF THIS BOOK MAY BE REPRODUCED TRANSMITTED, OR STORED IN ANY FORM OR BY ANY MEANS EXCEPT FOR YOUR OWN PERSONAL USE OR FOR A BOOK REVIEW, WITHOUT THE EXPRESS WRITTEN PERMISSION OF THE AUTHOR:

TINY CACTUS PUBLISHING

Help the spider, the ghost, the bats and the cats to meet on the pumpkin field to celebrate the Halloween.

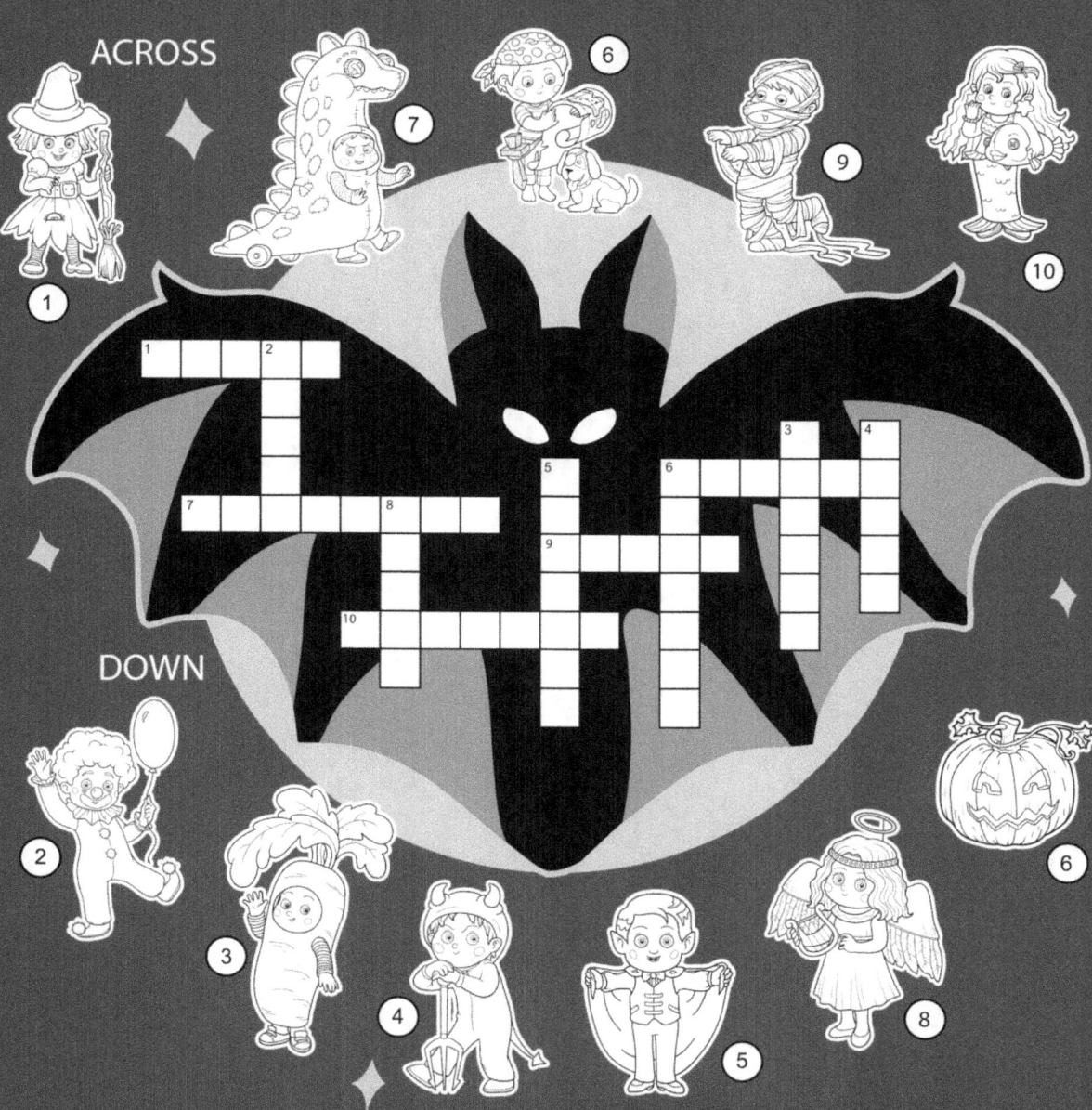

s			l		g			t
m			y		c			y
w			h		d			l

hos   and   umm   kul   evi   itc

Draw a line from dot number 1 to dot number 2, then from dot number 2 to dot number 3, 3 to 4, and so on. Continue to join the dots until you have connected all the numbered dots. Then color the picture!

Draw a line from dot number 1 to dot number 2, then from dot number 2 to dot number 3, 3 to 4, and so on. Continue to join the dots until you have connected all the numbered dots. Then color the picture!

www.ingramcontent.com/pod-product-compliance
Lightning Source LLC
Chambersburg PA
CBHW062200220526
45470CB00009B/2876